For Amy and Ryley
Hoping to make science fun!

ISBN-13: 978-0692953921
ISBN-10: 0692953922

1

One atom helps things fly

He
Helium Gas

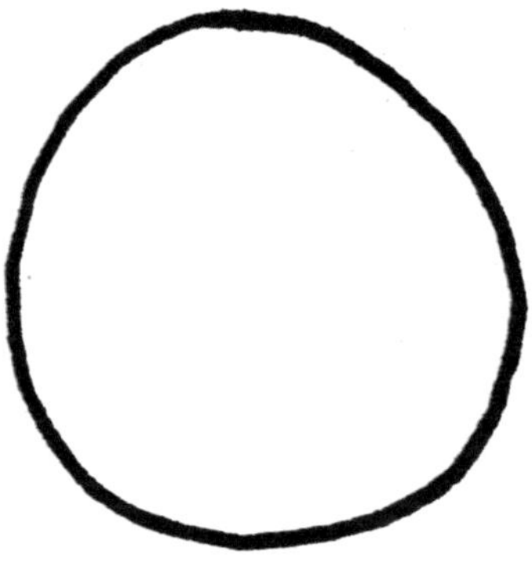

2

Two atoms fill the sky

N_2
Nitrogen Gas

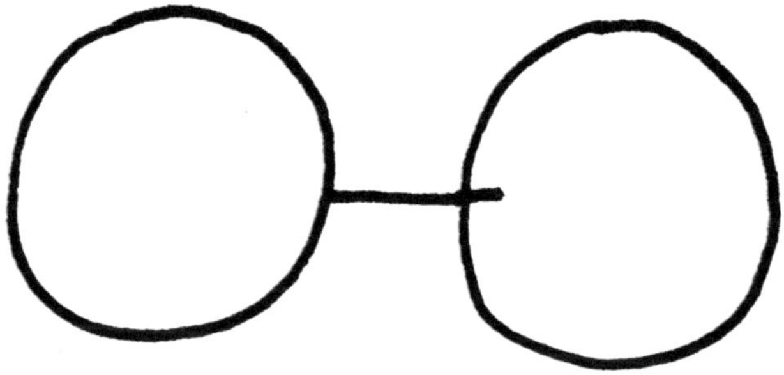

3

Three atoms fall as snow and rain

H_2O
Water

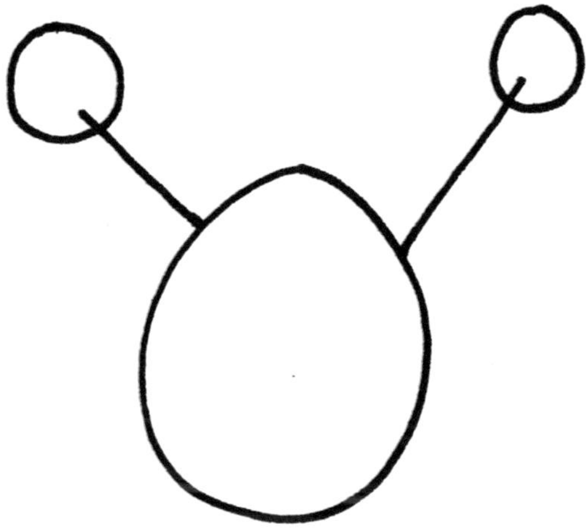

4

Four atoms clean the window panes

NH_3
Ammonia

5

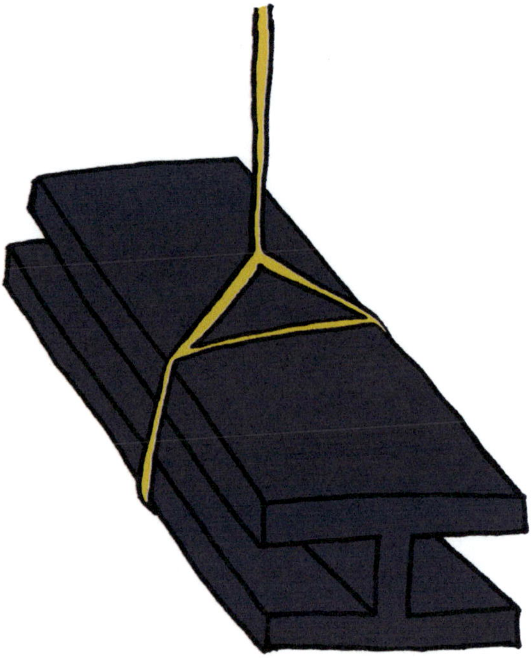

Five atoms are found in steel

Fe_2O_3
Ferric Oxide

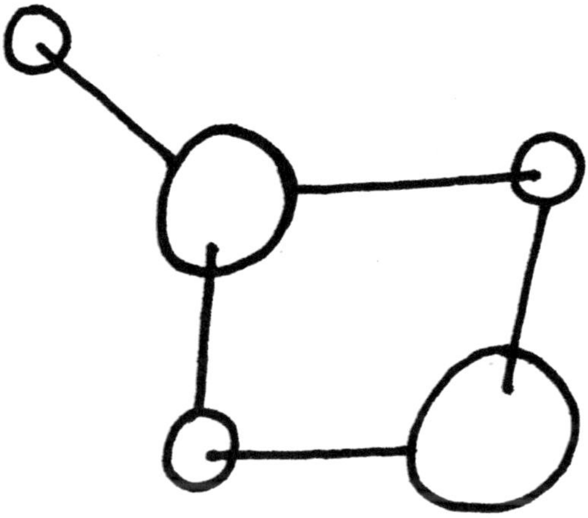

6

Six atoms help treat and heal

$MgSO_4$
Magnesium Sulfate

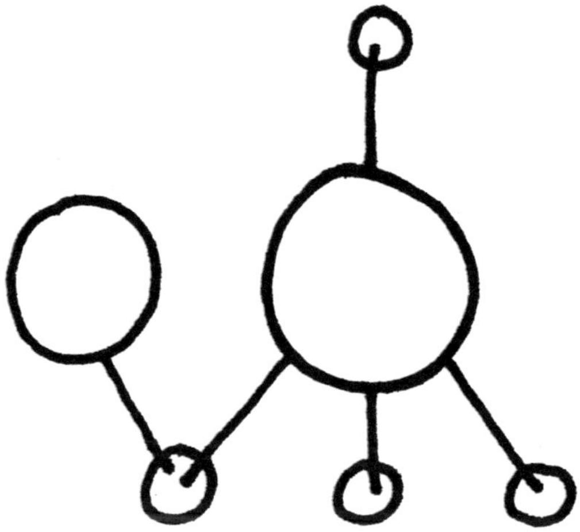

7

Seven atoms sparkle and shine

$MgAl_2O_4$
Magnesium Aluminate

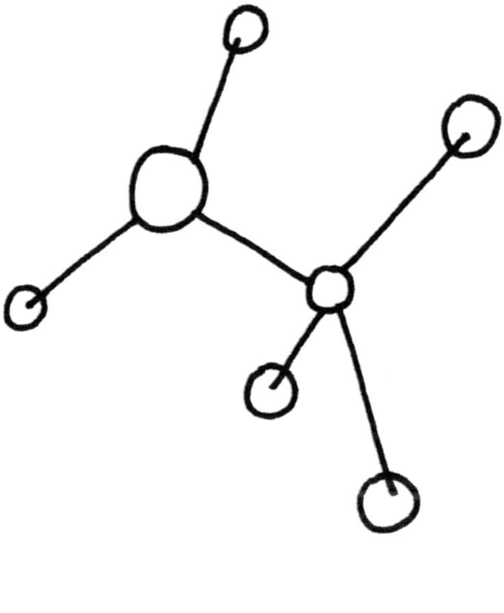

8

Eight atoms can cook, clean, and brine

CH_3COOH
Acetic Acid

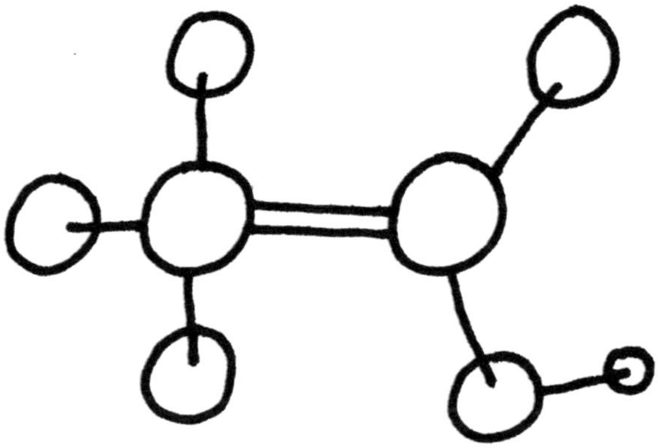

9

Nine atoms paint in blue

$C_3H_4O_2$
Acrylic Acid

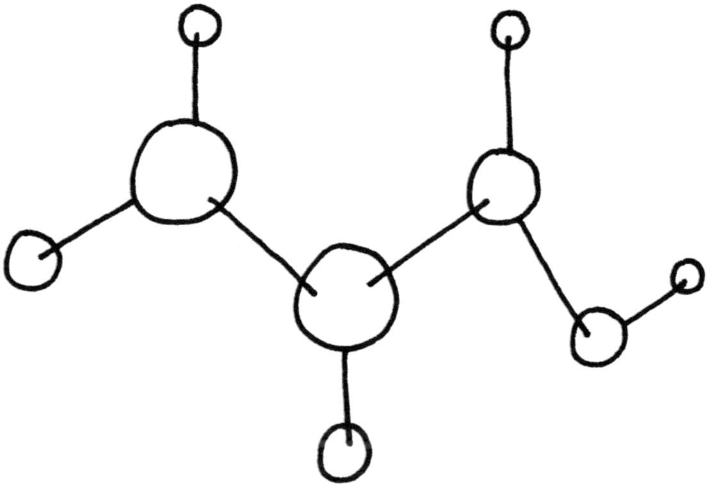

10

Ten atoms remove super glue

C_3H_6O
Acetone

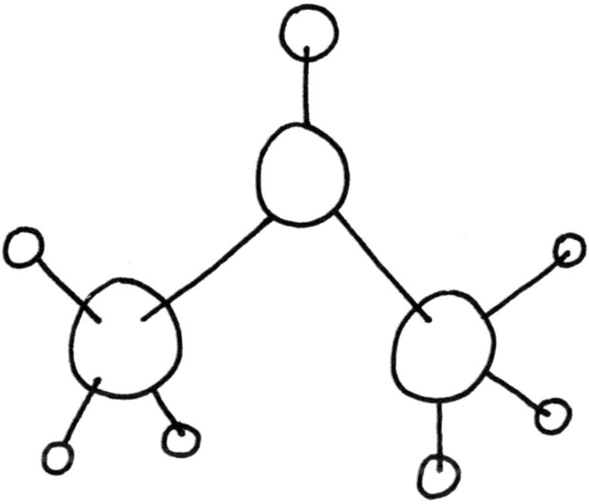

1 atom: He
Helium is a very noble gas.

2 atoms: N_2
Nitrogen is lighter than oxygen.

3 atoms: H_2O
Water expands when frozen.

4 atoms: NH_3
Ammonia is a frequently used cleaning agent.

5 atoms: Fe_2O_3
Ferric oxide is sometimes called rust.

6 atoms: $MgSO_4$
Magnesium sulfate can be used as bath salt.

7 atoms: $MgAl_2O_4$
Magnesium aluminate is found in gems and rubies.

8 atoms: CH_3COOH
Acetic acid has a sour taste.

9 atoms: $C_3H_4O_2$
Acrylic acid is common in paints and adhesives.

10 atoms: C_3H_6O
Acetone is a solvent often used in nail polish remover.

Made in the USA
Monee, IL
05 November 2026

67375662R00017